GREAT MARITIME INVENTIONS
1833-1950

GREAT MARITIME INVENTIONS

1833 1950

Mario Theriault

GOOSE LANE

Edited by Rhona Sawlor.
Book design by Julie Scriver.
Printed in Canada by Transcontinental Printing.
10 9 8 7 6 5 4 3 2 1

Canadian Cataloguing in Publication Data

Theriault, Mario, 1951-
Great Maritime inventions, 1833-1950

Includes bibliographical references and index.
ISBN 0-86492-324-4

1. Inventions — Maritime Provinces — History.
2. Patents — Maritime Provinces — History.
I. Title.

T227.M37T48 2001   608.7715   C2001-901729-4

Published with the financial support of the Canada Council for the Arts, the Government of Canada through the Book Publishing Industry Development Program, and the New Brunswick Culture and Sports Secretariat.

Goose Lane Editions
469 King Street
Fredericton, New Brunswick
CANADA E3B 1E5

To all my client inventors, who have inspired me to write this book,
and most of all to my wife Norma,
whose support and encouragement made all things easy.

ACKNOWLEDGMENTS

A note of appreciation is extended to the Public Archives of Nova Scotia, to the Public Archives and Records Office of Prince Edward Island, to the Provincial Archives of New Brunswick, to the Canadian Intellectual Property Office in Hull, and to the United States Patent and Trademark Office in Washington, DC. A special thank you is extended to Carole Choinière of the Canadian Intellectual Property Office and to Palmer C. DeMeo of Palmer Patent Consultants in Woodbridge, Virginia, for their valuable assistance in identifying these memorable inventions and locating the related patent documents. Thank you again to Mr. DeMeo for having researched and explained the US patent law in force during the period covered by this book. Thank you to Rhona Sawlor for having elegantly edited this book. Thank you to Cheryl Hulberg for her assistance in typing the manuscript. Thank you also to Yoland Mallet of the Canadian Intellectual Property Office for having rejuvenated a copy of the Canadian patent for the "Acme Club Skate," shown on page 12.

CONTENTS

13   **INTRODUCTION**

CONSUMER GOODS: FOOD

19   Ice Cream Soda
20   Confectionery Marker
21   Key-Opening Cans
22   Tea and Coffee Pot
23   Fish Oil Products
24   Fruit and Flake Cereal
26   Sardine Cans
27   Frozen Marinated Fish

CONSUMER GOODS: CONVENIENCES

28   Clothes Washer with Wringer Rolls
29   Clip-On Skates
30   Cold Water Soap
31   Ironing Board
32   Electric Telephony
33   Combined Hot and Cold Water Faucets
34   Vortex-Flushing Toilet Bowl
35   Adjustable Underwear Combination
36   Winter Cap with Foldable Elastic Ear
        and Head Band
37   Toothbrush with Replaceable Bristles
39   Gum Rubber Shoes
40   Replaceable Pool Cue Tip
41   Crossword Game

ENGINEERING: FARMING & INDUSTRY

45   Fishway

46   Scuba Tank

47   Compound Steam Engine

48   Kerosene

49   Potato Digger

50   Rubber-Lined Pulleys

51   Hay Carrier

53   Vane Pump

54   Roller Bearing

55   Connecting Link

56   Pulp Press and Wood Pulp in Sheets

57   Cultivating and Hilling Machine

ENGINEERING: TRANSPORTATION

58   Boat Propeller

60   Odometer

61   Snow Blower

62   Track Clearer (Miller's Flanger)

63   Separable Baggage Check

64   Cycle Runner

65   Wood and Canvas Canoe

66   Rein Connector (Quick-Release Buckle)

67   Airplane Position Indicator

68   Stabilizing Bar for Vehicle Suspension System

70   Automobile Backup Light

71   Dump Box for Truck

72   Variable Pitch Propeller

73   Partitioned Concrete Sidewalk

74   Red Pavement

75   Tempering of Steel Rails

ENGINEERING: CONSTRUCTION

76   Strain Gauges
77   Rotary Ventilator
78   Inserted Saw Tooth
79   Thermal Windowpane
80   Pipeless Furnace
82   Weatherstrip

83   **APPENDIX**

89   **BIBLIOGRAPHY**

91   **INDEX**

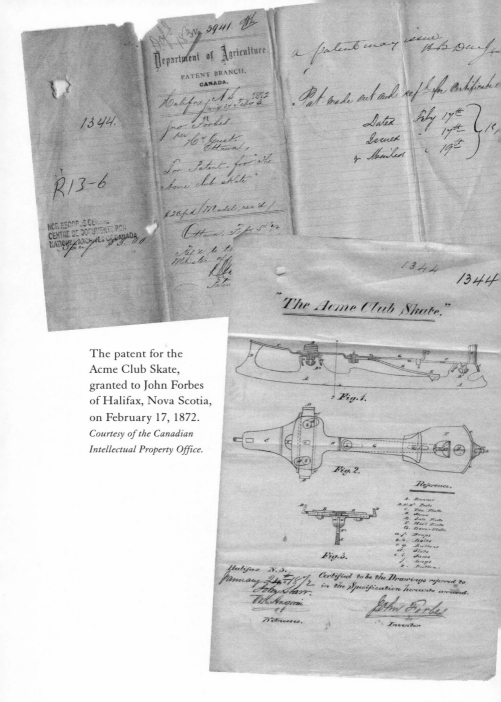

The patent for the
Acme Club Skate,
granted to John Forbes
of Halifax, Nova Scotia,
on February 17, 1872.
*Courtesy of the Canadian
Intellectual Property Office.*

INTRODUCTION

Letters Patent are granted to applicants for bringing into use new trades or new manufactures, for presenting new knowledge to the public, and for promoting the progress of science and useful arts. During the period covered by *Great Maritime Inventions, 1833-1950*, the two basic requirements for a patented invention were utility and novelty. (Inventiveness is a third requirement under modern patent law.) Concerning the element of utility, the invention had to be useful, to work as expected, and to produce the promised results. A model of the invention was often required to demonstrate utility. To meet the requirement of novelty, the invention must not have been known or used by others, in the province or in any other country, prior to its discovery.

Before Confederation, each province had its own patent system, modelled on the systems used in the United States and Great Britain. The following extract from the Revised Statutes of Nova Scotia, Chapter 120, entitled "Of Patents for Useful Inventions," dated 1833, outlines that province's first patent system:

> Whenever any person resident in the Province, and who shall have resided therein for the period of one year, or any British subject who shall have been an inhabitant of Canada, New Brunswick, Prince Edward's Island, or Newfoundland, for the space of one year previous to his application, shall apply to the Governor, alleging that he has discovered any new and useful art, machine, manufacture, or composition of matter, or any new and useful improvement thereon, not heretofore known or used, and pray that a patent may be granted to him for the same, the Governor may direct Letters Patent to be issued, reciting therein the allegations of such petition, and giving a short description of such

invention, and shall thereupon grant to the person so applying for the same and his representatives, for a term not exceeding fourteen years, the exclusive right of making, using, and vending the same to others, which Letters Patent shall be good and available to the grantee, and shall be recorded in the Secretary's Office, in a book for that purpose, and shall then be delivered to the patentee.

The 1833 statutes also stipulated: "Before any person shall obtain any Letters Patent, he shall make oath in writing that he verily believes that he is the true inventor or discoverer of the art, machine, or composition of matter, or improvement, for which he solicits Letters Patent, and that such invention or discovery has not been known in this province or any other country, which oath shall be delivered in with the petition for such Letters Patent."

A similar patent system was established in New Brunswick in 1834 and in Prince Edward Island in 1837. The patent systems of all three provinces were transferred to what is now the Canadian Intellectual Property Office in Hull shortly after Confederation, between 1869 and 1874.

The novelty or originality of the inventions described in *Great Maritime Inventions* is reasonably dependable. Each patent is either the first of its kind on record or the claims in it are so broad that the intellectual property protection afforded by the patent encompasses every possible precursor. The uniqueness of each invention is accepted based on the applicant's signed declaration of absolute novelty and on the knowledge and literature available to the examiners at the patent office at the time of the application.

Although great care has been taken to ensure that each invention is in fact a pioneer, it may be possible using modern information technologies to demonstrate that this or that invention was perhaps not quite the first of its kind in the world. On the other hand, it may also be possible to demonstrate that each alleged prior inventor had kept his invention as a trade secret, had disclosed it only to a select few under a confidentiality agreement, or simply did not file a patent application for it. Earlier inventions that are outside the patent system were also outside the scope of this book.

Under patent law, a patent is granted to an inventor who is willing to share his knowledge with the public. Each patent listed here went through a thorough examination by experts at the patent office. On the invention date, each invention was found to present new knowledge to the public and was seen as an important step forward in the evolution of science and the useful arts. Therefore, I considered the integrity of the patent office sufficient to support my claims of novelty.

Not all inventions are spectacular. In fact, most of the inventions described in *Great Maritime Inventions* never made it into the history books. Only after fifty years or more can we realize that every one of these inventions had an influence on our society. Each invention, large or small, was an important link in the chain of the evolution of science and the useful arts. Each invention has guided the progress of science in a certain direction, and each has improved to some degree the way we live.

The inventions are listed in chronological order within their categories. Each page contains an illustration and a short description of the invention, as well as the patent number and the name and address of the inventor(s). Copies of Canadian and US Patent documents are available to the public from the Canadian Intellectual Property Office (CIPO), Place du Portage I, 2nd floor, 50 Victoria Street, Hull, QC, Canada, K1A 0C9, by specifying the patent number and the inventor's name.

Great Maritime Inventions illustrates the importance of invention in our lives and instills pride in those who may recognize the genius of an ancestor. It describes some memorable inventions of residents of Nova Scotia, Prince Edward Island, and New Brunswick for which patents were granted between the introduction of the formal patent system and 1950; documents dated after 1950 may still be under copyright protection. During that period, over 3,300 patents were granted to residents of the Maritime Provinces.

The memorable inventions that have been selected for this book are those that mark great advances in science, those that substantially changed the course of development of technology, or those that have enjoyed a lasting success. Some of them are remembered or used to this day.

CONSUMER GOODS

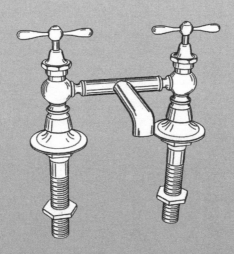

FOOD
CONVENIENCES

ICE CREAM SODA

INVENTOR: James William Black
Berwick, Nova Scotia
(Patent No. CA 24,012, July 5, 1886)

James William Black invented a syrup for making a refreshing beverage called Ice Cream Soda. The ingredients were whipped egg whites, sugar, water, lime juice, lemons, tartaric acid or citric acid, flavouring extract, and bicarbonate of soda. The ingredients were combined in this concentrated form and bottled. According to the inventor, a beverage of ice cream soda was readily mixed by placing a small ladle of the syrup in a drinking glass and filling the glass with ice water, which produced a beautiful drink, creamy and foaming at the top.

CONFECTIONERY MARKER

INVENTOR: Gilbert W. Ganong
St. Stephen, New Brunswick
(Patent No. CA 33,108, February 28, 1889)

As part of a marketing initiative, Gilbert W. Ganong of Ganong Bros. Ltd. invented a process for imprinting the letters GB on the bottom of individual chocolates. Each candy was dipped into melted chocolate, cream, or other confectionery and then placed upon the embossed or engraved letters or designs on a flexible mold plate and left there to dry. When hard, the chocolates were removed, with the design or letters of the mold plate now fully impressed upon them. The owners of Ganong Bros. Ltd. also invented the chicken bone, a hard candy jacket over a chocolate centre, in 1885. They invented the world's first chocolate nut bar in 1910. The company was the first in Canada to make lollipops (1895); the first in Canada to use cellophane in packaging (1920); the first to make chocolate peppermint rolls (1926); and the first in Canada to sell Valentine candy in heart-shaped boxes (1932).

FLEXIBLE MOLD PLATE

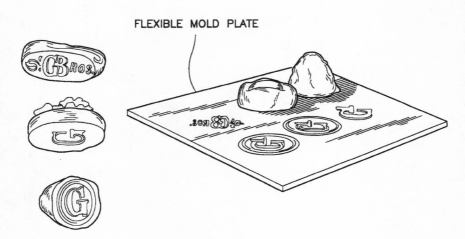

KEY-OPENING CANS

INVENTOR: Joseph Sutton Clark
St. George, New Brunswick
(Patent No. CA 68,455, August 13, 1900)

This invention is the earliest of the key-opening cans, in which the side of the can overlaps and ends in a lip or lug to which a key is attached to open the can. The inventor's disclosure also contains instructions for making flat cans. In that case, a seam is provided as close to the upper edge of the can as possible, in order to keep any liquid from spilling when the can is opened. Cans of this type are still used as containers for many foods, such as corned beef, ham, and other luncheon meats.

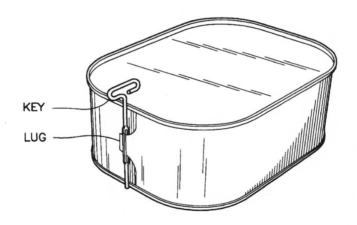

KEY

LUG

TEA AND COFFEE POT

INVENTOR: James Rooney
Kentville, Nova Scotia
(Patent No. CA 119,276, July 6, 1909)

This invention provided a perforated receptacle with a plunger inside a tea or coffee pot for containing tea leaves or coffee grounds, allowing them to be easily removed after infusion. A similar concept is still in use today.

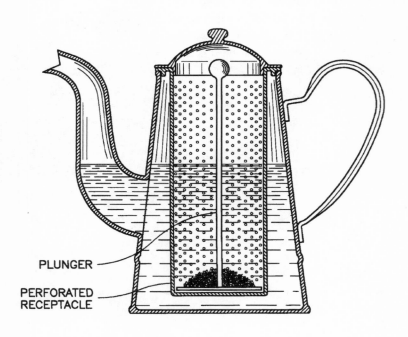

PLUNGER

PERFORATED
RECEPTACLE

FISH OIL PRODUCTS

INVENTOR: Hartley A. Wentworth

Deer Island, New Brunswick

(Patent Nos. CA 209,501, March 15, 1921; CA 333,890, July 11, 1933;
CA 378,373, December 13, 1938; CA 401,447, December 16, 1941)

Hartley A. Wentworth recognized the value of fish liver and fish oil as a food supplement, particularly for their high vitamin A and D content. He developed various processes for extracting and preparing fish liver, fish oils, and cod liver oil. Patent No. CA 333,890 illustrates Wentworth's commitment to improving nutrition. It describes a process for preparing a food supplement of fresh raw fish liver mixed with chocolate. This product was developed with the cooperation of another inventor from the area, Whidden Ganong of Ganong Bros. Ltd. Their objective was to make a food supplement that would be acceptable to children.

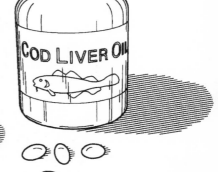

FRUIT AND FLAKE CEREAL

INVENTOR: George F. Humphrey
Bridgetown, Nova Scotia
(Patent Nos. CA 245,230, December 9, 1924; CA 275,794, November 29, 1927)

George F. Humphrey held the pioneer patents on breakfast foods containing fruits mixed with cereal products. He was the first to introduce into a breakfast food a natural fruit that imparted a distinct fruity flavour to the finished product and made it more nutritious.

Humphrey's first patent describes a method for producing a cold breakfast food. The fruits are cooked and reduced to pulp, and the pulp is dried, sweetened, and placed in a mixing and kneading machine. Cereal flour, such as wheat flour, is added to the fruit pulp, together with a solution of yeast and water, and the ingredients are mixed to create a uniform dough. The dough is cooked and passed through flaking rollers. The flakes are roasted in an oven, dried, cooled, and packed and sealed in containers for distribution. The second patent describes a process for producing a hot fruit and cereal breakfast food that would be cooked in the same manner as oatmeal. The process consisted of mixing an uncooked granulated cereal with an uncooked fruit pulp and then dehydrating the materials.

Humphrey's preferred fruit for the cold breakfast food was apples, because of their abundance and their nutritive value. He said, "The apple furnishes mineral salts and organic acids so necessary to the human system, and also vegetable albumens which promote and assist digestion." In the oatmeal-type breakfast food, Humphrey said, certain fruits could be used with certain cereals to create a variety of distinct flavours, such as oranges with corn meal and peaches with wheat semolina.

Because of George F. Humphrey's patents, breakfast foods such as Kellogg's Raisin Bran, Kellogg's Apple Crisp Müslix, and Kellogg's Banana Nut Müslix became obvious and thus unpatentable in a broad manner.

SARDINE CANS

INVENTOR: Henry T. Austin
Blacks Harbour, New Brunswick
(Patent No. CA 321,222, April 5, 1932)

The success of the sardine industry is attributable, to some degree, to the invention of a container that is a convenient size for carrying in a lunch box, holds enough sardines for a pleasant snack, and can be opened with a simple key that is detachable from the container. A number of inventors from southwestern New Brunswick, the home of Connors Bros. Ltd., the largest sardine plant of its kind in the world, contributed over time to the ultimate development of Henry T. Austin's sardine can, illustrated here. Their patents are listed on page 84.

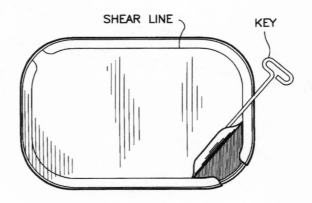

SHEAR LINE   KEY

FROZEN MARINATED FISH

INVENTORS: Walter H. Boutilier, Halifax, Nova Scotia,
and Frank W. Bryce, Montreal, Quebec
(Patent No. CA 386,404, January 23, 1940)

The frozen fish fillet was invented in Halifax around 1926. Fourteen years later, Walter H. Boutilier and Frank W. Bryce went one step further. Their patent describes a product consisting of raw fish impregnated with a flavouring substance, in which the fish is dipped prior to freezing. The patent also gave the inventors the exclusive right to the process of immersing fish in a low-temperature bath of a suitable flavouring substance and then subjecting it to quick freezing.

CLOTHES WASHER WITH WRINGER ROLLS

INVENTOR: John E. Turnbull
Saint John, New Brunswick
(Patent No. NB Archives, RS 549, Book H, page 329, July 10, 1843)

This inventor was the first to mount a set of wringer rolls on a washing machine. The rolls were activated by a crank and gearing system. The top roll was spring loaded to rise and fall with continuous pressure according to the thickness of the laundry passing between the rolls. The handle above the tub is part of a dashboard inside the tub. The handle was worked manually for ten minutes or so to wash each load of laundry.

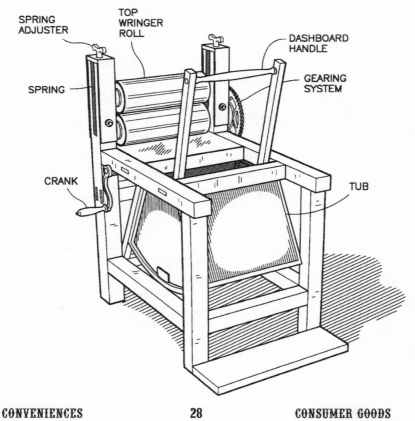

CLIP-ON SKATES

INVENTOR: John Forbes
Halifax, Nova Scotia
(Patent Nos. US 66,316, July 2, 1867; US 69,649, October 8, 1867;
CA 1,344, February 17, 1872; CA 1,348, February 22, 1872;
CA 1,423, April 23, 1872; CA 14,013, January 16, 1882;
CA 30,595, January 10, 1889; CA 30,706, January 19, 1889;
CA 46,523, June 13, 1894)

This invention is known as the world's first steel ice skate. It was manufactured by the Starr Manufacturing Company in Dartmouth, Nova Scotia, where John Forbes was foreman. The invention consisted of an all-steel skate blade which was clamped to the sole and heel of a boot by a single spring lever. This skate revolutionized skating and was instrumental in the development of hockey. It was promoted as the Acme Skate and was sold around the world. The Starr Manufacturing Company continued manufacturing this model of skate and various improved versions of it, some of which are described in the above patents, until 1938.

In fact, the steel skate is probably the one invention which has the most Maritime content — over 30 different kinds between 1867 and 1933. For a complete list of Maritime skate blade patents, see pages 85-86.

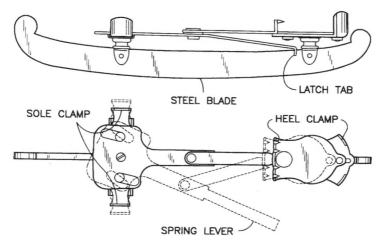

STEEL BLADE

LATCH TAB

SOLE CLAMP

HEEL CLAMP

SPRING LEVER

COLD WATER SOAP

INVENTOR: Andrew James Stewart
Saint John, New Brunswick
(Patent No. CA 1,175, October 4, 1871)

Andrew James Stewart saw a need for a soap which was more effective when used with cold, hard, or salt water than was ordinary soap, which was meant to be effective in warm soft water. So he invented a soap made with the ordinary fatty and alkaline ingredients of regular soap, but with the addition of a liquid concoction of borax dissolved in water, spirits of ammonia, naphtha, and spirits of turpentine. This soap would not likely be considered acceptable for sale or use under modern health and safety standards. However, it is believed that this invention was a valuable inspiration in the development of modern cold water soaps and detergents.

IRONING BOARD

INVENTOR: John B. Porter
Yarmouth, Nova Scotia
(Patent No. CA 4,653 April 6, 1875)

This patent describes two inventions. The first one consisted of an ironing board mounted on crossed legs which were hinged to one end of the board, so as to allow the legs to be folded underneath and the board to be stored out of the way when not in use. The second invention was a narrow press board for ironing sleeves and other small parts of clothing, fixed on the top of the ironing table at one end and secured in a clutch, the top of which was used as an iron stand. The press board was so arranged as to be removed when not needed. This second invention has since fallen out of use, but the first is still a standard design used in the manufacture of ironing boards.

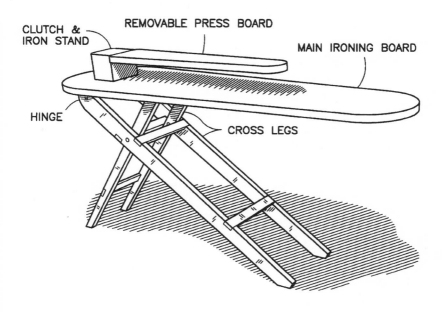

ELECTRIC TELEPHONY

INVENTOR: Alexander Graham Bell
Boston, Massachusetts; Baddeck, Nova Scotia, from 1885
(Patent Nos. CA 7,789, August 24, 1877; CA 10,705, November 27, 1879;
US 174,465, March 7, 1876; US 186,787, January 30, 1877;
US 201,488, March 19, 1878; US 213,090, March 11, 1879;
US 250,704, December 13, 1881)

The invention of the telephone was first described in the 1876 US patent and can be summarized as follows: the first armature is fastened loosely by one extremity to the leg of the first electromagnet, and its other extremity is attached to the centre of a stretched membrane. The first cone is used to converge sound vibrations upon the membrane. When a sound is uttered into this first cone, the membrane is set in vibration, the first armature is forced to take up the motion of the membrane, and thus electrical undulations are created upon the circuit. The undulatory current passing through the second electromagnet influences its armature to copy the motion of the first armature. A sound similar to that uttered in the first cone is then heard to proceed from the second cone. Although this circuit is admirable for its simplicity, the evolution of society that can be attributed to this invention is still progressing, with no limit in sight.

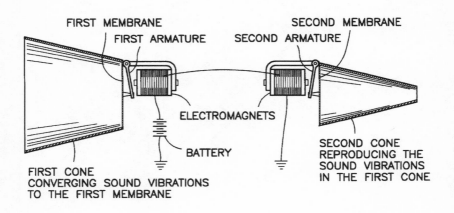

FIRST MEMBRANE

FIRST ARMATURE

SECOND MEMBRANE

SECOND ARMATURE

ELECTROMAGNETS

BATTERY

FIRST CONE
CONVERGING SOUND VIBRATIONS
TO THE FIRST MEMBRANE

SECOND CONE
REPRODUCING THE
SOUND VIBRATIONS
IN THE FIRST CONE

COMBINED HOT AND COLD WATER FAUCETS

INVENTOR: Thomas Campbell
Saint John, New Brunswick
(Patent No. CA 10,811, January 16, 1880)

It is difficult to imagine what our lives would be like without warm water at our shower heads, bathroom fixtures, kitchen sinks, and clothes washers. This invention made it possible and soon became an essential plumbing accessory worldwide. Although the invention seems simple in light of today's knowledge, the inventor was concerned that the faucet might not work because of back pressure. He solved this problem by making it clear in the patent disclosure that the discharge spout had to have a larger capacity than the combined discharges of both valves.

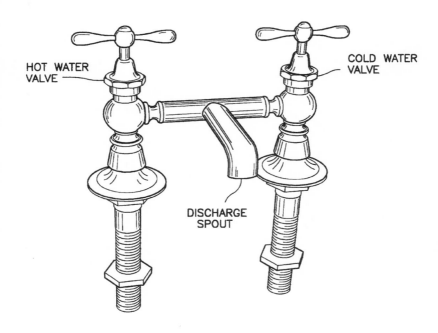

HOT WATER VALVE

COLD WATER VALVE

DISCHARGE SPOUT

VORTEX-FLUSHING TOILET BOWL

INVENTOR: Thomas McAvity Stewart
Saint John, New Brunswick
(Patent No. CA 108,017, October 15, 1907)

The patent for this invention describes a series of oblique openings in the rim of a toilet bowl to form a vortex during flushing, thus providing a self-cleansing effect. Today, most makes of toilet bowls have this feature or otherwise generate this same vortex effect.

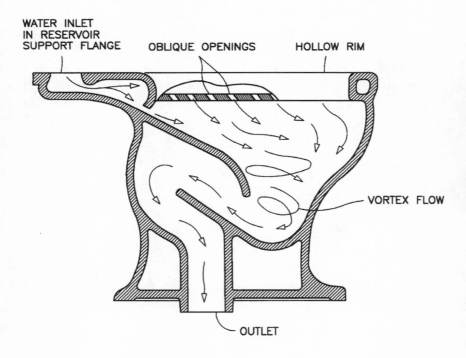

WATER INLET
IN RESERVOIR
SUPPORT FLANGE

OBLIQUE OPENINGS

HOLLOW RIM

VORTEX FLOW

OUTLET

ADJUSTABLE UNDERWEAR COMBINATION

INVENTOR: Frank Stanfield
Truro, Nova Scotia
(Patent No. CA 166,517, December 7, 1915)

The garment that many of us call "long johns" was originally known as "Underwear Combination" or "Combination Garment." The objectives of this invention were to provide long underwear in two separate pieces that could be readily adjusted to fit various trunk lengths and that could be worn separately or in combination as required. Further, the two pieces eliminated the need to wash one heavy single garment when, as Stanfield wrote in his patent disclosure, "often a part only requires cleansing." Prior to this invention, in 1898, Frank Stanfield and his brother John developed the famous Stanfield's Unshrinkable Underwear. Stanfield's Limited was incorporated in 1906 and remains a leader in quality undergarments.

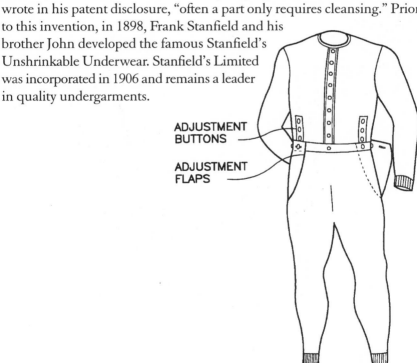

ADJUSTMENT
BUTTONS

ADJUSTMENT
FLAPS

WINTER CAP WITH FOLDABLE ELASTIC EAR AND HEAD BAND

INVENTOR: James C. Coulson

Truro, Nova Scotia

(Patent No. CA 175,717, March 13, 1917)

This fashionable men's cap had an inside banded flap adapted to be turned down to protect the ears and back of the head. The flap was made of woven cloth and was lined, and it had a knitted band on the outer edge. The fabric of the flap and the lining resisted the wind and kept the head effectively protected against the cold, and the elasticity of the knitted band kept the flap snug against the head. This type of cap has been in style ever since.

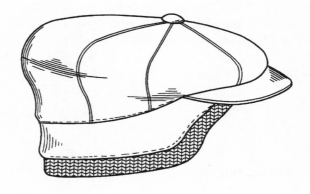

TOOTHBRUSH WITH REPLACEABLE BRISTLES

INVENTOR: Alfred C. Fuller
Hartford, Connecticut
(Patent No. CA 189,626, April 15, 1919)

Consideration for hygiene, efficiency, and economy can be seen in this novel toothbrush. The bristle assembly was cylindrical in shape to allow a rotary movement while brushing the teeth and therefore a more efficient cleaning. The bristle assembly was removable from the handle of the toothbrush, making it easy to clean the toothbrush after each use. As well, the bristle assembly could be replaced when worn. The bristle assembly was attached to the toothbrush head by a pivoting tab mounted at the base of the toothbrush head. This design illustrates well the concerns of this great entrepreneur for customer satisfaction.

Alfred C. Fuller was born in 1885 and raised in Welsford, King's County, Nova Scotia. He moved to Somerville, Massachusetts, in 1903, where he started the Fuller Brush Company in 1906. Fuller obtained at least two other patents for improvements to his toothbrush: patent nos. US 1,296,067, issued in 1919, and CA 222,949, issued in 1922.

The Fuller Brush Company grew quickly from a single man operation in 1906 to an enterprise having annual sales of $109 million in 1960. Today,

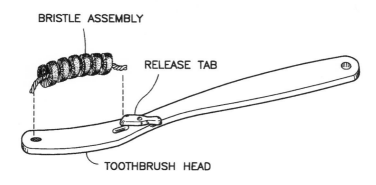

BRISTLE ASSEMBLY

RELEASE TAB

TOOTHBRUSH HEAD

the Fuller Brush Company has a twelve-acre manufacturing plant in Great Bend, Kansas, and produces more than 2,000 different products. During the period of strong growth, at least 125 US patents were granted to the Fuller Brush Company. This number of patents illustrates well that the company had a very prolific research and development department and that it believed in and encouraged innovation. For a complete listing of these patents, see page 87-88.

GUM RUBBER SHOES

INVENTOR: Charles L. Grant
Grand Pré, Nova Scotia
(Patent No. CA 199,520, April 27, 1920)

The patent for this invention describes a rubber shoe with the tongue as an integral part, making the shoe waterproof. The shoe was constructed with a rubber-coated tongue of a suitable shape. The tongue was vulcanized to the shoe during the operation of vulcanizing the shoe as a whole, making it impossible for the shoe to leak. This invention continues to be a standard practice in shoe-making.

REPLACEABLE POOL CUE TIP

INVENTOR: George W. Leadbetter
Springhill, Nova Scotia
(Patent No. CA 198,960, April 6, 1920)

This invention consisted in mounting a metal sleeve at the point of a billiard cue and threading a cue tip into the sleeve. The tip is thus easily replaced when necessary. In modern pool cues, the metal sleeve has been replaced by a plastic sleeve, and the tip is glued on. However, it is believed that this invention has influenced to a substantial extent the construction of modern pool cues.

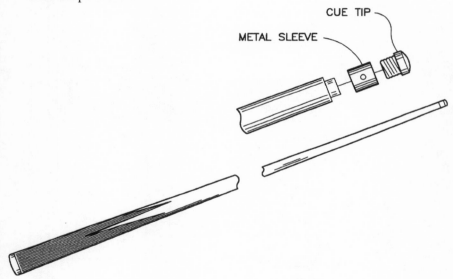

CROSSWORD GAME

INVENTOR: Edward R. McDonald
Shediac, New Brunswick
(Patent No. CA 266,459, December 7, 1926)

This invention consisted of a game played on a checkerboard with square game pieces that were individually lettered to represent the alphabet. Each letter-carrying game piece, with the exception of the vowels, a, e, i, o, and u, also had a number indicating the value of the letter on the piece. The game was played by two players who each had a complete set of lettered game pieces, each set of a different colour. Play began with the players arranging their letters in two rows on each side of the checkerboard in the squares provided. The object of the game was for each player to form a word using as many as possible of the letters in his set or "stealing" from his opponent's set as required. The players took turns, moving one letter out from its

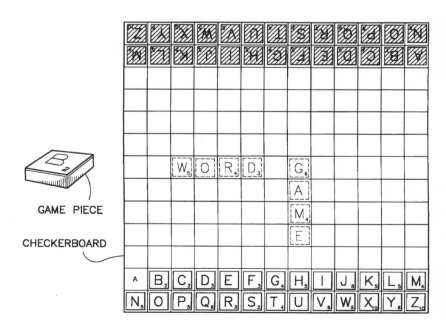

GAME PIECE

CHECKERBOARD

original position to any unoccupied square on the board. It is believed that this crossword game was the precursor of the well-known Scrabble game. Scrabble is a registered trademark of Hasbro Inc. Research by the National Scrabble Association of USA indicates that Scrabble was invented between 1930 and 1933. The finding of this Canadian patent pushes back the invention to September 28, 1925, the application date of this patent.

ENGINEERING

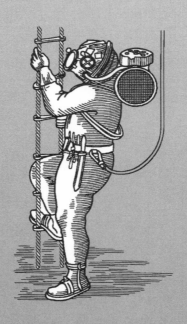

FARMING & INDUSTRY
TRANSPORTATION
CONSTRUCTION

FISHWAY

INVENTOR: Richard McFarlan
Bathurst, New Brunswick
(Patent No. NB Archives, RS 549, page 19, July 3, 1837)

Richard McFarlan invented a fishway around a mill dam, allowing fish to ascend and descend the river in the same manner as they had before the dam was built. The fishway was constructed with a series of step-like ponds with connecting underwater passages, which the fish could swim through to get around the dam. These passages were designed to maintain a certain level of water inside the ponds under normal river conditions while maintaining a certain flow of water to mill machinery.

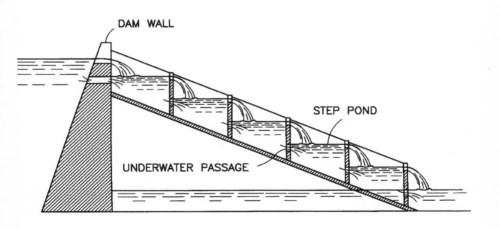

SCUBA TANK

INVENTORS: James Elliott and Alexander McAvity
Saint John, New Brunswick
(Patent: NB Archives, RS 549, page 9, March 4, 1839)

By patenting this "oxygen reservoir for divers," the inventors made public their design for a piece of equipment that has remained essential to both amateur and professional divers for more than 160 years. Its modern name is an acronym: a self-contained underwater breathing apparatus, a scuba tank. Before Confederation, each province had its own patent system, and so this patent was granted by the Lieutenant Governor of New Brunswick.

The inventor wrote, "Instead of supplying air in that manner [with a force pump], the individual going under water carries with him a quantity of condensed oxygen gas or common atmospheric air proportionate to the

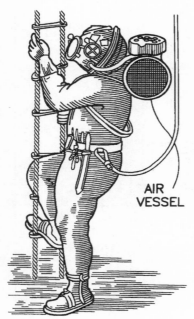

AIR
VESSEL

depth of water and adequate for the time he is intended to remain below. This gas or air to be contained in vessels of any size made of copper or other suitable metal, and used in any requisite quantity by means of a pipe, regulated by stop cock, leading from the vessel in which it is contained and communicating with the interior of the dress, and which dress the inventors contemplate using either for the whole or part of the body. Another valve is placed at the top of the head, through which the foul air escapes, and thus in a great measure preserves pure the air which is inhaled by the operator under water."

COMPOUND STEAM ENGINE

INVENTOR: Benjamin F. Tibbets
Fredericton, New Brunswick
(Patent Nos. CA 85, November 6, 1845, CA 411, July 2, 1853)

Benjamin F. Tibbets was the first in Canada to use the steam discharged from a primary cylinder of a marine steam engine in a second cylinder, thereby increasing the efficiency of the engine. His invention called for the use of a reservoir large enough to contain the steam discharged from the high-pressure cylinder. The reservoir then supplied a second low-pressure cylinder with steam of sufficient density to operate it efficiently. The original engine built by Tibbets remained in use in various ships for 70 years.

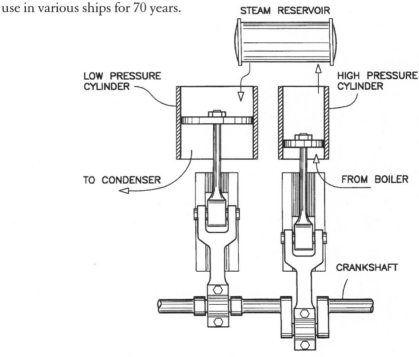

KEROSENE

INVENTOR: Abraham Gesner
Halifax, Nova Scotia
(Patent Nos. US 11,203, US 11,204 and US 11,205, all issued on June 27, 1854)

Abraham Gesner seems to have been ahead of his time when he invented kerosene. Concerned about fuel consumption, he wrote in his patent application: "As the rocks whence the kerosene is most abundantly obtained are widely disseminated, the deposits of them are of almost unlimited extent, an immense mass of hitherto useless matter will by means of this invention be rendered available for the uses of mankind as a cheap and convenient substitute for illuminating purposes for the oils and fats which are yearly increasing in scarcity and price."

Gesner's method of obtaining fuel from either petroleum, maltha or soft mineral pitch, asphaltum, or bitumen involved dry distillation, treatment with reagents, and redistillation. The process resulted in three different types of flammable liquids, which Gesner labelled Kerosene A, Kerosene B, and Kerosene C, each of which had different properties and uses. Kerosene A was highly volatile and flammable but produced good light. Kerosene B was only moderately volatile and flammable but did not produce as good a light as did Kerosene A. Kerosene C, the heaviest liquid of the three, burned well in a certain type of lamp and was a good lubricant for machinery. Abraham Gesner's inventions have often been referred to as the precursors of all modern petroleum fuels.

POTATO DIGGER

INVENTORS: William Bradford Allin and William Stiggins
Charlottetown, Prince Edward Island
(Patent No. PEI Archives, RG 7, Series 2, File 9, pages 1-13, September 2, 1868)

Modern potato harvesting machines and potato handling equipment owe their beginning to Allin and Stiggins's invention of the endless chain elevator. It was operated by a gearing system mounted onto the wheels of the machine. The wheels had projections to grip the ground and transmit the torque required to turn the gearing system. The potatoes were raised by the endless chain elevator from a scoop at the front end of the machine and into a sifter at the rear end, thus removing the soil from the potatoes as they were harvested.

A forerunner of this machine was invented by another Prince Edward Islander. A cylindrical elevator with flat teeth to separate the potatoes from the earth was invented by George Jenkins, Township 49, Queen's County, PEI (Archives, RG 7, Series 2, File 15, pages 122-126, August 1, 1860).

Patents for major improvements to Allin and Stiggins's machine were later granted to other residents of Prince Edward Island as follows: patent no. 10,066 to John Solomon Cantelo, Grand River, PEI, June 11, 1879; and patent no. 21,999 to W.E. Reynolds, Murray Harbour, PEI, July 4, 1885.

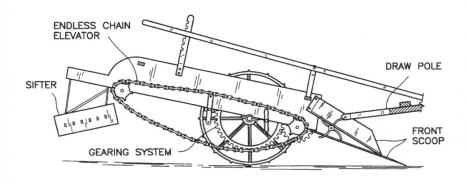

RUBBER-LINED PULLEYS

INVENTOR: Isaac McNaughton
Egerton, Pictou County, Nova Scotia
(Patent No. CA 1,451, May 7, 1872)

The object of this invention was to prevent endless belts or metal cables from slipping on shaft pulleys and sheaves during operation, thus prolonging their wear. This invention consisted in wrapping and bonding a band of vulcanized rubber around the periphery of pulleys and sheaves . Since this concept was introduced, the rubber lining of pulleys and sheaves has been standard practice in the manufacture of equipment used for power transmission.

FLAT PULLEY   RUBBER LINING

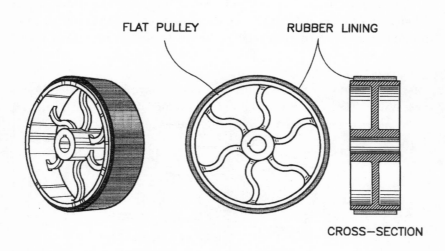

CROSS—SECTION

HAY CARRIER
INVENTOR: Abraham Gill the Younger
Township 34, Prince Edward Island
(Patent No. PEI Archives, RG 7, Series 2, File 7, pages 199-203, October 8, 1872)

This patent describes a beam-and-trolley-type carrier mounted under a barn roof for unloading hay at various places in a barn's loft. Although hay is now stored in bales, this type of carrier can still be found in older barns and was still commonly used into the 1970s. The patent document describes the invention as follows: "A hay carrier which consists of a long beam say three inches thick, nine inches deep, and from forty to fifty feet long suspended edgeways immediately under the ridge of the roof, and also of a carriage to run on the upper edge of this long beam. The carriage consists of a horizontal piece of wood about two feet long, three and one-quarter inches thick or square with an iron sheave at each end to run on an iron round bar on the top edge of the long beam. There is a seven-inch sheave

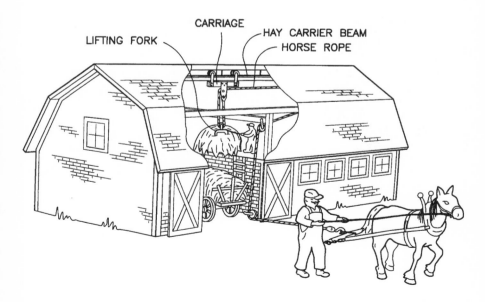

CARRIAGE

LIFTING FORK

HAY CARRIER BEAM
HORSE ROPE

suspended from the centre of the carriage for the horse rope to work in under the main beam. It is worked by a rope of sufficient strength with an American Lifting Fork at one end of the main rope and a single sheave at the end of the main beam in the building, another single sheave fixed at or near the lower floor and a horse attached for working at the other end of the main rope. As soon as the cross head of the fork rides to the large sheave under the running carriage, the carriage then moves to the place desired to deposit the content of the fork; the fork being cleared of its content by the use of a small cord attached to the lever of the fork and used by the person in the building."

VANE PUMP

INVENTOR: Charles C. Barnes
Sackville, New Brunswick
(Patent No. CA 3,559, June 16, 1874)

The invention consisted of a wheel with diametrical sliding leaves which revolved in a casing that had a segmental enlargement. During operation, this formed a suction and pressure chamber that communicated with inlet and outlet openings. When the wheel rotated, the leaves slid in and out against the casing, causing a continuous suction through the inlet opening and a pressure flow through the outlet opening. A check valve was installed on the inlet pipe to resist back pressure and to keep the inlet pipe full. The same design is still used today.

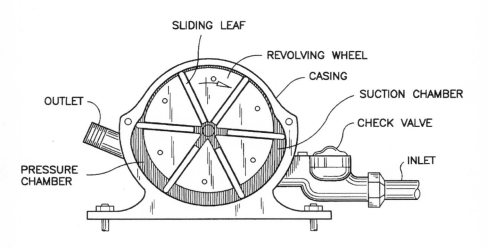

ROLLER BEARING

INVENTOR: George Welton Thomas
Bear River, Digby County, Nova Scotia
(Patent No. CA 9,843, April 12, 1879)

George Welton Thomas came up with an innovation that would prove to be an invaluable contribution to modern machine design. His objective was to design a way of assembling machinery so that the friction generated by the contact of moving parts with stationary ones (such as wheel hubs and their axles) would be greatly reduced, if not done away with entirely. The patent for this invention describes the combination of rollers and a roller cage with the shafts and housings of milling and factory machinery, vehicles, railway cars, and farming equipment.

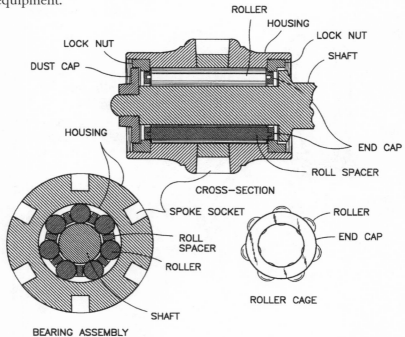

CONNECTING LINK

INVENTOR: Donald Munro
Pictou, Nova Scotia
(Patent No. CA 21,779, May 30, 1885)

This invention consists of an open link, threaded on each end, with a sleeve that has an interior thread. By turning the sleeve in one direction, the opening in the link is exposed, and by turning it in the opposite direction, the opening can be closed. These connecting links can still be found, as first conceived by the inventor, in most hardware stores.

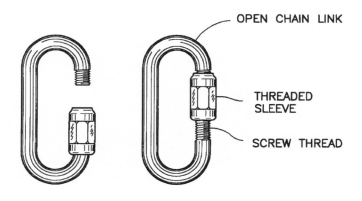

OPEN CHAIN LINK

THREADED SLEEVE

SCREW THREAD

PULP PRESS AND WOOD PULP IN SHEETS

INVENTOR: Joseph Stuart Hughes
Chesley Corner, Lunenberg County, Nova Scotia
(Patent Nos. CA 52,680, June 17, 1896; CA 77,058, August 8, 1902;
CA 77,210, August 26, 1902; and CA 77,211, August 26, 1902)

Not all pulp and paper mills manufacture a finished paper product. It is common for a mill near abundant wood resources to manufacture pulp as an intermediate product. This pulp is generally sold to other paper mills located in large cities far from a wood supply. Wood pulp of various types is also used to make blends to control specific paper characteristics, such as fibre length, strength, brightness, etc. The above patents relate to the packaging, storing, and transporting of wood pulp as an intermediate product. The first three patents describe pulp presses. The exclusive property granted to the inventor by the fourth patent is defined as "a new commercial product, consisting of uncut rectangular sheets, plain cakes or slabs of wood pulp in a dry or semi-dry state." Joseph Stuart Hughes was the first to conceive of handling wood pulp in dry rectangular sheets. Since then, the drying and baling of pulp sheets has been a standard practice in the paper industry around the world for manufacturing, selling, storing, and transporting wood pulp.

PULP BALE

CULTIVATING AND HILLING MACHINE

INVENTOR: Robert C. Holman
Summerside, Prince Edward Island
(Patent No. CA 341,416, May 8, 1934)

This invention combined scuffler teeth or tines with mould-boards or discs in a single implement, so that the earth could be thoroughly cultivated and hilled into rows in one operation. This concept has been in use ever since.

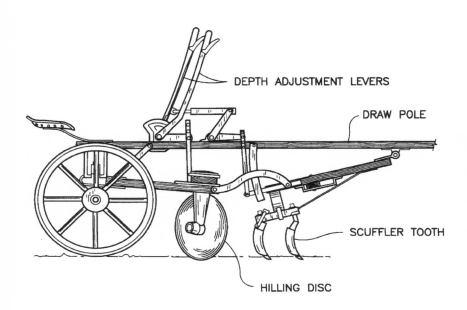

DEPTH ADJUSTMENT LEVERS

DRAW POLE

SCUFFLER TOOTH

HILLING DISC

BOAT PROPELLER

INVENTOR: John Patch
Yarmouth, Nova Scotia
(Patent No. US 6,914, November 27, 1849)

The signed affidavits of 75 residents of Yarmouth, Nova Scotia, support the fact that John Patch invented a boat propeller in 1833 and demonstrated it in a small boat in the harbour of that town. Patch described his first propeller as "an instrument acting in the water on the principle of an oar in sculling a boat." The patent system in Nova Scotia was only in its infancy, so Patch chose to go to Washington, DC, to apply for a patent.

His application was denied because Patch did not meet the residency requirement of the American patent system. Disappointed but undaunted, he returned home and continued to work on improvements to his propeller. Sixteen years later, he invented a second propeller, which he called the "double-action propeller." This time, the US patent shown here was granted, as residency restrictions had been modified and Patch now had a Boston address on his application. This second invention did not receive much attention from the press, as other inventors in Europe and in the United Sates were then making headlines with other types of propellers

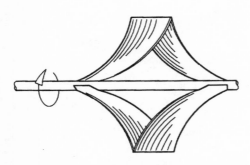

PATCH'S DOUBLE–ACTION PROPELLER

MODERN
BOAT PROPELLER

that operated very differently. Time has proven that John Patch's concept was more practical and efficient than the other types. The modern propeller (shown on page 58) is surprisingly similar to a single pair of the blades in Patch's double action propeller and fits beautifully the original definition of Patch's first invention: "an instrument acting in the water on the principle of an oar in sculling a boat."

ODOMETER

INVENTOR: Samuel McKeen,
Mabou, Nova Scotia
(Patent No. NS Archives, RG 5, Series GP, Vol. 13, item 25, March 31, 1854)

Those who have seen the inside of a modern odometer will realize that, aside from the size and the arrangement of the gears inside the counter mechanisms, it has a remarkable similarity to McKeen's invention. This early odometer consisted of a series of gear plates mounted on the frame of a horse-drawn carriage to keep track of the distance travelled. The carriage wheel had a circumference of 16 feet and 6 inches, travelling one rod per complete rotation. The first gear plate engaged with a pinion on the hub of the carriage wheel. The gear ratio between the pinion and the first gear plate was 8:1, so that the first gear plate had a full rotation in eight rods. Similar gearing systems were used to rotate the other gear plates. The second gear plate had a full rotation in 80 rods, 1/4 mile. The third gear plate had a full rotation in two miles, and so on, up to the sixth gear plate. An indicator hand encircled by a dial was placed upon the appropriate gear plates to mark off the rods and miles travelled. A hammer and bell were attached to strike off the miles as the carriage rolled along.

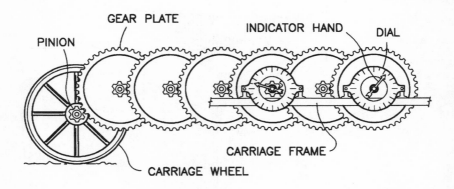

SNOW BLOWER

INVENTOR: Robert Carr Harris
Dalhousie, New Brunswick
(Patent No. CA 594, September 5, 1870)

Many claim to be the inventor of the snow blower. However, Robert C. Harris seems to have been the first to conceive of a snow removing machine that combined a screw auger and a fan-type blower. He called it the "Railway Screw Snow Excavator." The screw auger was mounted vertically and fed snow from a scoop area to the blower mounted horizontally above the auger. The machine was mounted on the front of a locomotive and was powered by the engine of the locomotive by way of a shaft, a sheave, and a gearbox set above the blower. This invention is believed to be the precursor of modern snow blowers.

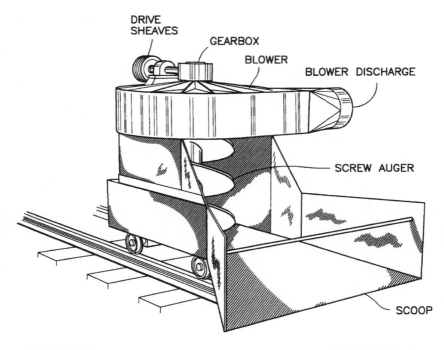

DRIVE SHEAVES

GEARBOX

BLOWER

BLOWER DISCHARGE

SCREW AUGER

SCOOP

TRACK CLEARER (MILLER'S FLANGER)

INVENTOR: James H. Miller
Fredericton, New Brunswick
(Patent Nos. CA 2,391, May 28, 1873; US 138,913, May 13, 1873;
US 178,687, June 13, 1876)

James H. Miller was concerned that ice and snow build-up on railway tracks caused loss of power and delays in the running of trains. So he invented a pair of rail scrapers that were attached to the cowcatcher of a locomotive. The metal plates were raised or lowered by the engineer, using levers and rods that extended from the scrapers back to the cab of the engine. This successful innovation remained popular with railway companies for many years.

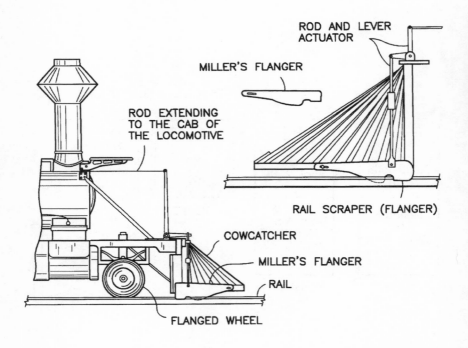

ROD AND LEVER ACTUATOR

MILLER'S FLANGER

ROD EXTENDING TO THE CAB OF THE LOCOMOTIVE

RAIL SCRAPER (FLANGER)

COWCATCHER

MILLER'S FLANGER

RAIL

FLANGED WHEEL

SEPARABLE BAGGAGE CHECK

INVENTOR: John Mitchell Lyons
Moncton, New Brunswick
(Patent No. CA 14,911, June 5, 1882)

This invention consisted of a separable coupon ticket, showing on both halves the name of the issuing station, the destination, and a consecutive number, and a wood or metal holder. The coupon half was torn off and given to the owner of the baggage. The check half was secured to the holder by a strap which was attached to the piece of baggage. Modern baggage checks are not inserted in a brass holder, but otherwise the original concept is still used in bus, train, and plane travel.

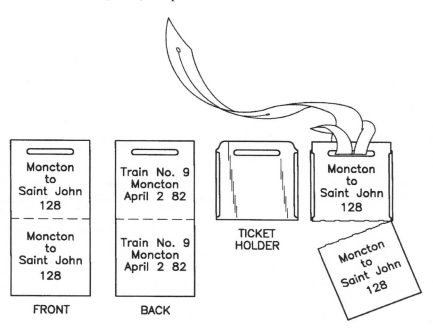

CYCLE RUNNER

INVENTOR: J. Roswell Sederquest
St. Stephen, New Brunswick
(Patent No. CA 45,592, March 20, 1894)

At a time when the bicycle was a primary mode of transportation, this invention must have been a great success. It provided a way of getting around in winter that was as easy, safe, and speedy as bicycling in summer. This cycle was easy to propel in snow, with a front runner mounted in place of the front wheel. According to the inventor, the runner made it possible to pedal the cycle with not more than half the power needed to pedal a two-wheeler through snow. The main advantage of the runner was that, in snow two or three inches deep, it cleared a track for the rear wheel and made travelling easy where it would have been impossible with the two wheels attached. An improved version of this winter bicycle is still marketed worldwide under the trademark Snowbike by Corimba S.N.C., an Italian corporation.

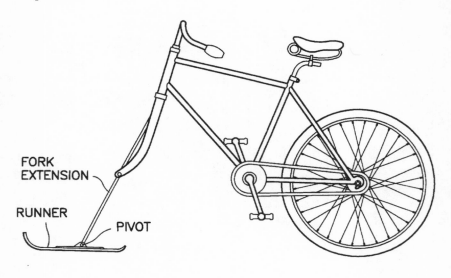

FORK
EXTENSION

RUNNER

PIVOT

WOOD AND CANVAS CANOE

INVENTOR: William T. Chestnut
Fredericton, New Brunswick
(Patent Nos. CA 91, 848, February 23, 1905; CA 93,181, May 11, 1905;
CA 117,374, March 23, 1909)

The Chestnut Canoe Company was the pride of Fredericton for over 80 years (1897-1979). The company maintained an average employment of 60 craftsmen and contributed greatly to the local economy. Because of the first patent listed above, the inventor and his brother Harry obtained the exclusive right in Canada to manufacture and sell their famous wood and canvas canoes. Chestnut soon became, and still remains, a symbol of quality and excellence in canoes; by 1915, the Hudson's Bay Company was buying all the canoes the company could make, and the Chestnut Canoe Company became the largest canoe manufacturer in Canada. The manufacturing of these canoes was described in the first patent document as "bending a plurality of transverse ribs about a suitable wooden form, securing the ends of the ribs to a suitable gunwale, covering the ribs with a plurality of longitudinal slats, covering the slats with a covering of canvas, and treating the canvas cover with an impervious filling material."

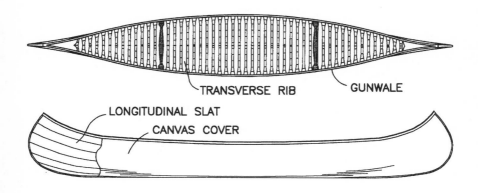

TRANSVERSE RIB   GUNWALE

LONGITUDINAL SLAT

CANVAS COVER

REIN CONNECTOR
(QUICK-RELEASE BUCKLE)

INVENTOR: Arthur Davy
New Glasgow, Nova Scotia
(Patent No. CA 134,438, May 18, 1911)

When Arthur Davy came up with this invention, he probably did not imagine that it would be used, some 90 years later, on articles of clothing, school bags, packsacks, car seats, harnesses, and carry cases of all sorts. This quick-release buckle was invented as a connector for horses' reins. The inventor said: "The objects of my invention are to provide a simplified and effective form of device for connecting the extremities of two reins to replace the ordinary buckle, and such a device as will permit the ready and rapid attachment and detachment of the extremities of the reins. My invention consists essentially of two members adapted to be connected to the extremities of the rein, one member of which carries a pair of spring actuated dogs adapted to extend into and engage slots in the other member. It will be seen that when connected there are no protruding parts on the device likely to catch on anything during movement of the rein. The device is extremely simple in construction and will be found to possess considerable advantages over the ordinary buckle. When once locked, it securely retains the ends of the reins together, and detachment can only be made by pressing the dogs inwardly."

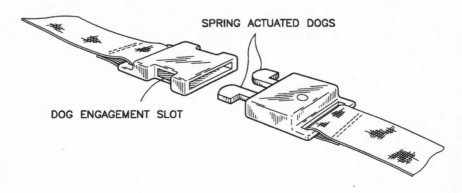

SPRING ACTUATED DOGS

DOG ENGAGEMENT SLOT

AIRPLANE POSITION INDICATOR

INVENTOR: John D. MacGillivray
Antigonish, Nova Scotia
(Patent No. CA 192,304, August 26, 1919)

John D. MacGillivray's invention of the position indicator was an important step in the development of modern aircraft instrumentation. He designed a gauge that would indicate the precise attitude of the plane when flying above the clouds. The gauge was affixed to the frame of the airplane, in sight of the pilot. It consisted of a pair of balls placed inside a cylindrical cage. The positions of the balls along the cylindrical cage and against one of its four bars indicated whether the plane was flying upright, upside down, tilted to one side, climbing, or descending.

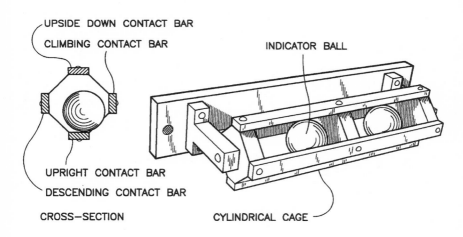

UPSIDE DOWN CONTACT BAR
CLIMBING CONTACT BAR
INDICATOR BALL
UPRIGHT CONTACT BAR
DESCENDING CONTACT BAR
CROSS—SECTION
CYLINDRICAL CAGE

STABILIZING BAR
FOR VEHICLE SUSPENSION SYSTEM

INVENTOR: Stephen Leonard Chauncey Coleman,
Fredericton, New Brunswick
(Patent No. CA 189,894, April 22, 1919)

S.L.C. Coleman was the first to propose the use of stabilizer bars on the suspensions of motor vehicles. This patent describes the equalizer bars attached to the vehicle's springs and to pivot brackets on its front and rear. This allowed the equalizer bars to work with the springs to reduce the lateral rolling of the vehicle when driving on rough surfaces and around curves. The stabilizing bars on modern vehicles are of simpler construction but are nonetheless still very similar to the original concept.

Stephen Leonard Chauncey Coleman was a prolific inventor whose ideas added to the advancement of automotive technology. In a business proposal to Lord Beaverbrook, in 1949, he described himself as an engineer, and he

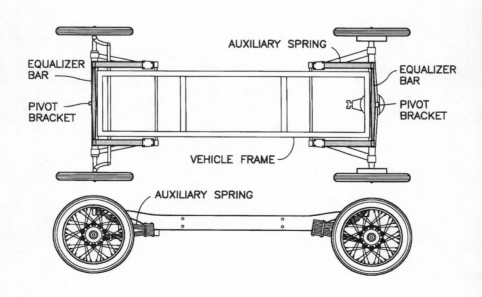

wrote, "For the past 35 years I have engaged in inventing and patenting improvements in the construction of automobiles, and was ably assisted by the late Professor John Stevens of UNB [the University of New Brunswick], who was a mechanical genius." For a list of some of Coleman's patents, along with those of other Maritimers who contributed in various ways to the automotive field, see pages 83-84.

AUTOMOBILE BACKUP LIGHT

INVENTOR: James A. Ross
Halifax, Nova Scotia
(Patent No. CA 195,537, December 30, 1919)

Early automobiles came equipped with headlights at the front, of course, but these were of no use for observing the roadway at the rear of the vehicle. Aware that nighttime driving was a concern for many, especially when one had to back out of a narrow passageway or turn the vehicle around, James A. Ross came up with a bright idea. He mounted a light at the rear of the vehicle and connected it to a switch affixed to the base of the gearshift lever. The switch activated the light when the gearshift lever was placed in the reverse position. This invention has been in use ever since.

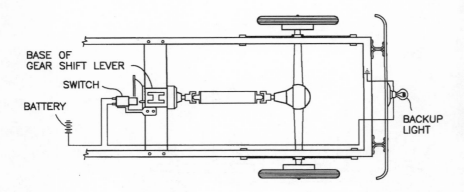

DUMP BOX FOR TRUCK

INVENTOR: Robert T. Mawhinney
Saint John, New Brunswick
(Patent No. CA 203,004, August 17, 1920)

This invention was instrumental in the development of our present trucking industry. To create this first dump truck, a mast was mounted between the cab of the vehicle and the dump box. A cable was threaded over a sheave at the top of the mast and was connected to a winch at the base of the mast and to the lower front end of the dump box. The dump box was pivoted at the rear end of the truck frame. A simple crank handle was used to operate the winch, which raised the front end of the dump box, dumped the load, then lowered the box. A hydraulic system has since replaced the crank handle, but the basic concept has remained unchanged.

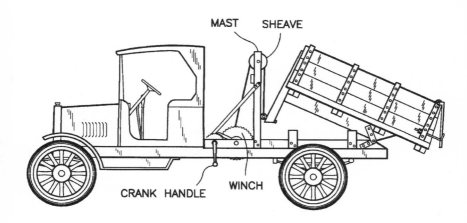

VARIABLE PITCH PROPELLER

INVENTOR: Wallace Rupert Turnbull
Rothesay, New Brunswick
(Patent Nos. CA 215,654, February 7, 1922; CA 280,497, May 29, 1928;
CA 281,602, July 10, 1928)

This invention provided a mechanism for varying the pitch of propeller blades to suit flying conditions during the operation of an aircraft. For example, the pitch had to be different for gaining altitude than for flying straight ahead. The variable pitch propeller also allowed the pitch of the blades to be reversed to create a braking effect, allowing for landing in shortened distances. In this sense, the variable pitch propeller does for the airplane what the transmission does for the car. Before this invention, the airplane ran in one gear all the time. It could fly but could not carry a reasonable payload. The variable pitch propeller made it possible for airplanes to get off the ground with larger payloads and carry them to distant destinations. For these reasons, this invention may well be one of the most important developments in the history of aviation.

BLADE JOINT

ADJUSTMENT MECHANISM COVER   ADJUSTABLE PORTION OF BLADE

PARTITIONED CONCRETE SIDEWALK

INVENTORS: Arthur Wesley Hall and William Alexander McVay
St. Stephen, New Brunswick
(Patent No. CA 239,600, April 29, 1924)

Before Hall and McVay's innovation, sidewalks were made of one continuous strip of concrete that was vulnerable to the cracking and heaving actions of frost. The invention consisted of incorporating soft joints of sand or other loose filling material at intervals along the sidewalk, forming a series of juxtaposed slabs. The soft joints allowed each slab to work with the frost without breaking. The inventors foresaw that, with the disappearance of the frost, the slabs would settle back into their original positions, leaving the sidewalk with a level surface and the slabs intact. Thus, at a minimum cost, they had invented a sidewalk that was practically immune to the effects of the frost. To this day, this invention remains standard practice in building sidewalks.

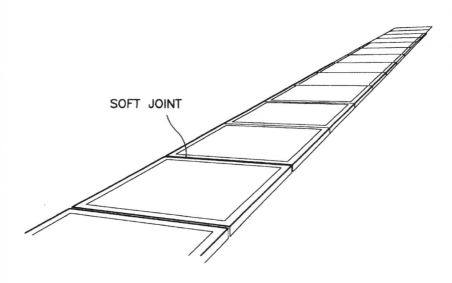

SOFT JOINT

TEMPERING OF STEEL RAILS

INVENTOR: Irwin Cameron Mackie
Sydney, Nova Scotia
(Patent No. CA 319,553, February 22, 1932)

Irwin Cameron Mackie, the chief metallurgist for the Dominion Steel and Coal Corporation of Sydney, Nova Scotia, developed a process to eliminate shatter cracks in the steel rails of train tracks. Cracks had always been a serious problem, often causing the rails to fail in service. This inventive method for manufacturing rails consisted principally of slowing the rate of cooling of the newly milled rails over a specific temperature range. By the1940s, most rail producers worldwide were using the Mackie cooling method. In recognition of his accomplishment, Mackie was granted honourary membership in the Canadian Standards Association.

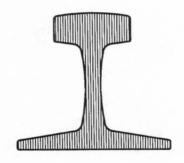

RED PAVEMENT
INVENTOR: Henry James Phillips
Charlottetown, Prince Edward Island
(Patent No. CA 321,794, April 26, 1932)

Henry James Phillips was a Prince Edward Island road builder with considerable experience in making pavement. After years of experimenting, he discovered an ideal pavement made with regular asphalt and a mixture of the unique red clay and red sandstone found in the subsoil of Prince Edward Island. Apparently, this was the first time any type of clay was used in an asphalt pavement. Red pavement prepared according to this invention can still be seen on some of the Island's roads and in southeast New Brunswick.

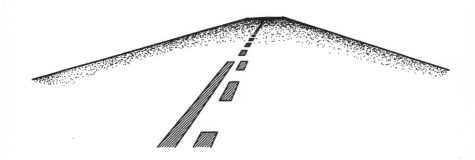

STRAIN GAUGES

Inventor: John Forbes
Dartmouth, Nova Scotia
(Patent No. CA 7,870, September 7, 1877)

John Forbes's invention was an important innovation that led to the design of modern strain gauges and their use in monitoring static structures, such as roofs and bridges. This invention consisted of attaching an insulated wire to each member of a structure so that an electrical circuit would be made or broken by any abnormal condition resulting from excessive strain or dislocation of parts of the structure. The closing or rupture of the circuit was indicated by an alarm. In demonstrating his invention, Forbes selected a structure having three members in compression — the upper beams forming the arc — and three members in tension — the two vertical rods and the lower horizontal beam. In modern strain gauges, the electrical resistance in a semiconductor strip is measured and translated to indicate strain in the strip.

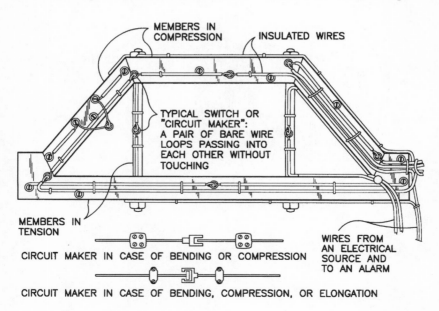

MEMBERS IN COMPRESSION

INSULATED WIRES

TYPICAL SWITCH OR "CIRCUIT MAKER": A PAIR OF BARE WIRE LOOPS PASSING INTO EACH OTHER WITHOUT TOUCHING

MEMBERS IN TENSION

CIRCUIT MAKER IN CASE OF BENDING OR COMPRESSION

WIRES FROM AN ELECTRICAL SOURCE AND TO AN ALARM

CIRCUIT MAKER IN CASE OF BENDING, COMPRESSION, OR ELONGATION

ROTARY VENTILATOR

Inventor: James Thomas Lipsett
Saint John, New Brunswick
(Patent No. CA 30,854, February 14, 1889)

This type of ventilator, still popular for use on residential air exchanger systems, uses wind power to increase the updraft in a chimney or roof vent. The blades are designed to carry melting snow, ice, or rain to the outside of the ventilator so that moisture will not drip down inside the chimney or the ventilation conduit and rust it.

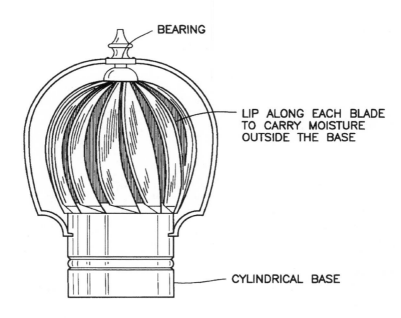

BEARING

LIP ALONG EACH BLADE
TO CARRY MOISTURE
OUTSIDE THE BASE

CYLINDRICAL BASE

INSERTED SAW TOOTH

Inventor: Philias Bertrand
Saint John, New Brunswick
(Patent No. CA 48,989, May 20, 1895)

Although there have been many different types of replaceable saw teeth, this invention continues to be the preferred model in the sawmill industry. It was described as a tooth formed in two parts, one of the parts being the "bit," which has the cutting edge of the tooth, while the other part, the "key," holds the bit in its place in the saw. A portion of the key part is arranged to give a spring pressure against the bit, securing the bit against any rattling or working loose. When the two parts are placed together, they create a circular outline. The outer edge of the tooth is grooved to fit over the V-shaped edge of the circular gap in the saw plate.

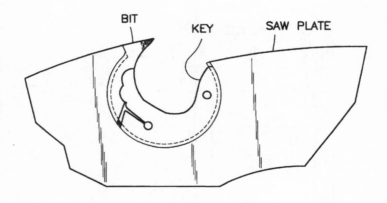

THERMAL WINDOWPANE

Inventor: Lawrence St. Clair McCloskey
Boiestown, New Brunswick
(Patent No. CA 175,060, February 13, 1917)

This patent describes a window having two glass panes separated by a hollow space, much the same as modern thermal windows. The edges of the panes were hermetically sealed, and the space between was filled with alcohol. The inventor wrote: "By use of the alcohol, it has been found that the light rays can freely pass through the window, and that the furnishings of the room fitted with a window of this type do not suffer in consequence of their being exposed to the sunlight, while at the same time the temperature of the room is better controlled, being warmer in winter and cooler in summer."

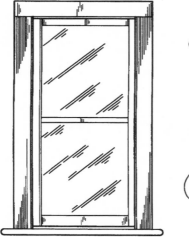

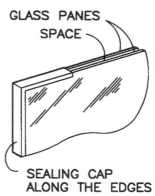

GLASS PANES
SPACE

SEALING CAP
ALONG THE EDGES

PIPELESS FURNACE

Inventor: L.W. Daman
Sackville, New Brunswick
(Patent No. CA 239,975, May 13, 1924)

This type of furnace was characterized by its large air diffuser, which was set into the floor of a central room on the main floor of the building, directly above the furnace. The metal grill over the heat diffuser was often used to dry mittens and boot liners on wet days. Many a child lost marbles and crayons through the openings of the large floor grill. The objects of this invention were: to eliminate pipes and tubes within the heater; to insure the delivery of the heated air beyond the furnace room; to provide good air

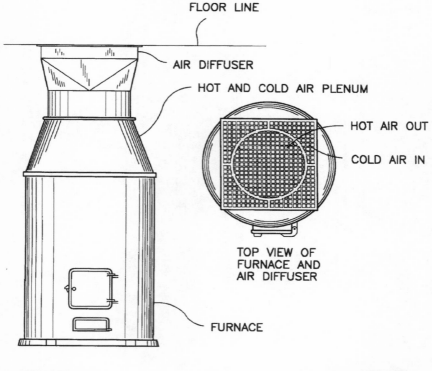

FLOOR LINE

AIR DIFFUSER

HOT AND COLD AIR PLENUM

HOT AIR OUT

COLD AIR IN

TOP VIEW OF
FURNACE AND
AIR DIFFUSER

FURNACE

circulation; to facilitate assembly, insure convenience, and reduce cost in shipping and packaging; and generally to provide a safe, durable, and efficient heating furnace. This type of furnace was popular until the 1970s, although many units can still be found heating older homes today.

WEATHERSTRIP

Inventor: Joseph Leon Therriault
Saint Hilaire, New Brunswick
(Patent No. CA 417,156, December 21, 1943)

This new weatherstrip was developed with the idea that it would not contain rubber or metal because these two materials were in restricted supply during the war. The innovative design has a wood molding with a cushion strip, made of felt, cotton, or other flexible material, attached to one edge. The molding has a series of slots to hold the screws that attach it to a frame. These slots allow the molding to move against the frame when the screws are slackened slightly, making the weatherstrip adjustable. This type of weatherstrip can still be found in hardware stores.

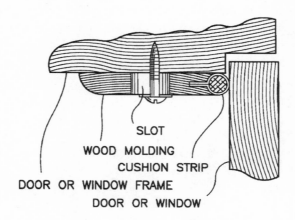

SLOT
WOOD MOLDING
CUSHION STRIP
DOOR OR WINDOW FRAME
DOOR OR WINDOW

APPENDIX
Major Contributions to Various Technologies
by Maritime Inventors

During the preparation of this book, I was greatly impressed with the work of some prolific inventors and their contributions to the evolution of various technologies. Some of these inventions were not first in a specific field or were not outstanding successes. However, I have listed some of these to show that the work of one inventor often inspires another, and so on, and that this is the way science and technology progress.

AUTOMOBILE PARTS

The following inventors contributed to the development of vehicle brakes, shock absorbers, suspension systems, universal joints, air springs, ball joints, rotary engines, and wheel rims. These innovations are illustrated and described in the following patents:

1905	CA 95,874	Joseph Thompson, Westfield, New Brunswick
1918	CA 188,848	John M. Spencer, Truro, Nova Scotia
1919	CA 189,894	Stephen Leonard Chauncey Coleman, Fredericton, New Brunswick
1919	CA 189,895	Stephen Leonard Chauncey Coleman, Fredericton, New Brunswick
1919	CA 189,896	Stephen Leonard Chauncey Coleman, Fredericton, New Brunswick
1919	CA 189,897	Stephen Leonard Chauncey Coleman, Fredericton, New Brunswick
1922	CA 225,750	John Henry Poole, Saint John, New Brunswick
1929	CA 295,354	Stephen Leonard Chauncey Coleman, Fredericton, New Brunswick
1935	CA 350,681	Rupert Leslie Rand, Amherst, Nova Scotia
1940	CA 387,497	Chesley Ernest Smith, Amherst, Nova Scotia

1942	CA 402,416	Stephen Leonard Chauncey Coleman, Fredericton, New Brunswick
1944	CA 422,291	Stephen Leonard Chauncey Coleman, Fredericton, New Brunswick
1947	CA 439,119	Stephen Leonard Chauncey Coleman, Fredericton, New Brunswick
1950	CA 464,743	Stephen Leonard Chauncey Coleman, Fredericton, New Brunswick

GARMENTS

Charles Lewis of Truro, Nova Scotia, contributed to the development of head wear, diapers, shirt collars, shirt cuffs, and the preparation of mixtures of wool and nylon fibres for use in the manufacture of blended textile yarns and fabrics. Such innovations are described in his patents:

1935	CA 354,354	1936	CA 356,221	1940	CA 387,823
1936	CA 356,220	1938	CA 376,028		

SARDINE CANS

The following patents describe different versions of the key-opening containers:

1913	CA 147,415	Francis Partridge McColl, St. Andrews, New Brunswick
1913	CA 147,416	Francis Partridge McColl, St. Andrews, New Brunswick
1913	CA 147,417	Francis Partridge McColl, St. Andrews, New Brunswick
1914	CA 158,388	Francis Partridge McColl, St. Andrews, New Brunswick
1915	CA 160,085	Francis Partridge McColl, St. Andrews, New Brunswick
1917	CA 177,754	Francis Partridge McColl, St. Andrews, New Brunswick
1918	CA 184,284	Lewis Connors, Blacks Harbour, New Brunswick
1932	CA 321,222	Henry T. Austin, Blacks Harbour, New Brunswick

STEEL SKATES

This list of patents illustrates the Maritime contributions to the development of the steel skate:

1867	US 66,316	John Forbes, Halifax, Nova Scotia
1867	US 69,649	John Forbes, Halifax, Nova Scotia
1869	CA 180	Edward Lawson Fenerty, Halifax, Nova Scotia
1870	CA 568	William Henry Barker, Windsor, Nova Scotia
1871	CA 992	Bernard Gallagher, Saint John, New Brunswick
1871	CA 1,026	John T. Larkin, Halifax, Nova Scotia
1872	CA 1,344	John Forbes, Halifax, Nova Scotia
1872	CA 1,348	John Forbes, Halifax, Nova Scotia
1872	CA 1,423	John Forbes, Halifax, Nova Scotia
1872	CA 1,579	Edward Lawson Fenerty, Halifax, Nova Scotia
1874	CA 3,122	James Albert Whelpley, Dartmouth, Nova Scotia
1874	CA 3,788	Edward Lawson Fenerty, Halifax, Nova Scotia
1876	CA 5,694	Edward Lawson Fenerty, Halifax, Nova Scotia
1877	CA 7,922	Samuel Horsford, Halifax, Nova Scotia
1877	CA 8,460	Robert Gay, Dartmouth, Nova Scotia
1882	CA 14,013	John Forbes, Halifax, Nova Scotia
1888	CA 30,595	John Forbes, Halifax, Nova Scotia
1889	CA 30,706	John Forbes, Halifax, Nova Scotia
1890	CA 35,402	George Charles Bateman, Halifax, Nova Scotia
1893	CA 43,036	Edward Lawson Fenerty, Halifax Nova Scotia
1893	CA 44,513	Richard Daine, Halifax, Nova Scotia
1893	CA 44,929	Thomas Harrison, Dartmouth, Nova Scotia
1894	CA 46,523	John Forbes, Halifax, Nova Scotia
1894	CA 46,886	Thomas Harrison and Edwin H. Whelpley, Dartmouth, Nova Scotia
1895	CA 49,074	Thomas Harrison, Dartmouth, Nova Scotia
1898	CA 59,556	Robert Bustin, Saint John, New Brunswick
1898	CA 61,365	Thomas Harrison and Henry Goudge, Dartmouth, Nova Scotia.
1901	CA 74,307	Robert Bustin, Saint John. New Brunswick.
1909	CA 119,838	Rachel A. Thompson, Oxford, Nova Scotia
1919	CA 192,403	W.A. Crowell, Dartmouth, Nova Scotia

1921	CA 213,687	Sidney Herbert Goodenough, Dartmouth, Nova Scotia
1925	CA 252,639	John R. Robertson, Dartmouth, Nova Scotia
1933	CA 336,740	Peter Taylor, Dartmouth, Nova Scotia

TIDAL MOTORS

The harnessing of tidal and wave power challenged many inventors in the past and continues to be investigated as a potential source of renewable energy. The following documents describe various interesting concepts and constitute valuable background material to inspire inventors in the future:

1906	CA 100,174	George Whitman, Round Hill, Annapolis County, Nova Scotia
1906	CA 102,692	Frank A. Harrison, Sackville, New Brunswick
1914	CA 159,088	W[allace] R[upert] Turnbull, Rothesay, New Brunswick
1921	CA 212,430	Osborne H. Parson, Halifax, Nova Scotia
1922	CA 217,833	Osborne H. Parson, Halifax, Nova Scotia
1923	CA 233,477	Osborne H. Parson, Halifax, Nova Scotia
1933	CA 336,888	Osborne H. Parson, Halifax, Nova Scotia

TYPEWRITER AND CALCULATOR MECHANISMS

The following patents describe improvements to calculating machines, adding machines, addressing machines, and typewriters invented by Frank A. Harrison of Sackville, New Brunswick:

| 1907 | CA 105,028 | 1906 | CA 112,979 | 1907 | CA 120,276 |
| 1906 | CA 109,832 | 1906 | CA 120,275 | | |

WATER PURIFICATION AND METAL PROTECTION

Through the following inventions, Frank Negus of Halifax, Nova Scotia, made major contributions to the advancement of science in the fields of electrolytic purification of water and the cathodic protection of the metal of ships and boilers:

1944	CA 420,178	1947	CA 441,657	1948	CA 446,079
1946	CA 436,815	1947	CA 441,658		
1947	CA 441,656	1947	CA 442,170		

HOUSEHOLD CLEANING PRODUCTS

Following is the complete list of US and Canadian patents granted to Alfred C. Fuller, founder of the Fuller Brush Company, and to his employees:

1913	US 1,058,203	1927	US 1,630,282	1929	US 1,725,133
1919	CA 189,626	1927	US 1,635,970	1929	US 1,729,977
1919	CA 191,591	1927	US 1,635,971	1929	US 1,734,503
1919	US 1,296,067	1927	US 1,636,186	1930	US 1,743,319
1921	US 1,389,302	1927	US 1,638,074	1930	US 1,749,732
1922	CA 222,949	1927	US 1,646,052	1930	US 1,749,733
1922	US 1,435,641	1927	US 1,646,571	1930	US 1,749,737
1923	US D 62,726	1927	US 1,650,248	1930	US 1,749,744
1923	US D 63,074	1928	US D75,320	1930	US 1,760,268
1923	US 1,446,599	1928	US D 75,688	1930	US 1,761,180
1924	US 1,479,109	1928	US 1,655,085	1930	US 1,763,738
1924	US D66,289	1928	US 1,659,707	1930	US 1,763,836
1924	US 1,504,129	1928	US 1,662,777	1930	US 1,764,095
1924	US 1,504,147	1928	US 1,667,911	1931	US 1,789,120
1924	US 1,504,148	1928	US 1,671,930	1931	US 1,795,159
1924	US 1,513,556	1928	US 1,681,322	1931	US 1,805,219
1924	US 1,519,335	1928	US 1,689,287	1931	US 1,806,519
1924	US 1,519,359	1928	US 1,691,094	1931	US 1,806,520
1925	US 1,526,568	1928	US 1,691,159	1931	US 1,807,559
1925	US 1,527,188	1928	US 1,692,110	1931	US 1,834,911
1925	US 1,534,231	1928	US 1,693,263	1932	US 1,845,209
1925	US 1,537,918	1928	US 1,694,955	1932	US 1,850,853
1925	US 1,547,611	1929	US 1,701,088	1932	US 1,851,537
1925	US 1,557,655	1929	US 1,703,086	1932	US 1,856,026
1926	US 1,569,431	1929	US 1,706,714	1932	US 1,871,775
1926	US 1,587,093	1929	US 1,706,715	1933	US 1,897,968
1926	US 1,588,810	1929	US 1,710,661	1933	US 1,911,550
1926	US 1,588,940	1929	US 1,711,412	1935	US 2,017,829
1926	US 1,598,119	1929	US 1,711,461	1937	US 2,070,393
1926	US 1,600,258	1929	US 1,716,844	1937	US 2,095,917
1927	US 1,620,803	1929	US 1,719,093	1937	US 2,095,918
1927	US 1,628,615	1929	US 1,724,453	1937	US 2,101,799

1939	US 2,146,624	1942	US 2,295,933	1944	US 2,341,728
1939	US 2,152,429	1942	US 2,295,970	1944	US 2,356,121
1939	US 2,176,861	1942	US 2,299,709	1944	US 2,358,443
1940	US 2,223,147	1943	US 2,308,674	1945	US 2,367,650
1941	US 2,230,968	1943	US 2,310,011	1945	US 2,374,415
1941	US 2,267,584	1943	US 2,310,897	1945	US 2,378,553
1942	US 2,271,835	1943	US 2,312,591	1945	US 2,380,310
1942	US 2,272,598	1943	US 2,313,037	1945	US 2,380,311
1942	US 2,281,412	1943	US 2,314,306	1946	US 2,400,809
1942	US 2,289,313	1943	US 2,329,434	1946	US 2,402,333
1942	US 2,290,534	1943	US 2,332,490		

BIBLIOGRAPHY

Bourne, John C.E. *A Treatise on the Screw Propeller with Various Suggestions of Improvement*. London: Longman, Brown, Green and Longmans, 1855.

Brown, J.J. *Ideas in Exile: A History of Canadian Inventions*. Toronto: McClelland and Stewart, 1967.

Carpenter, Thomas. *Inventors: Profiles in Canadian Genius*. Camden East ON: Camden House, 1990.

Coleman, Stephen Leonard Chauncey. Letter to Lord Beaverbrook. University of New Brunswick, Harriet Irving Library, Case 52, File 3, Item 31934, 1949.

Dobson, Susan, Cathy Simon and Suzanne Wood. *New Brunswick Firsts*. New Brunswick Bicentennial, Provincial Archives of New Brunswick reference no. MC80/1065, 1984.

Eber, Dorothy Harley. *Genius at Work: Images of Alexander Graham Bell*. Halifax: Nimbus, 1991.

Folster, David. *The Chocolate Ganongs of St. Stephen, New Brunswick*. Toronto: Macmillan, 1990; Fredericton: Goose Lane, 1991.

Fox, Harold G. *The Canadian Law and Practice Relating to Letters Patent for Inventions*. Toronto: Carswell, 1969.

Fuller, Alfred C. *A Foot in the Door*. New York: McGraw-Hill, 1960.

Gardiner, Robert. *The Advent of Steam*. London: Conway Maritime, 1993.

Mayer, Roy. *Inventing Canada: One Hundred Years of Innovation*. Vancouver: Raincoast, 1997.

Mitcham, Allison. *The Best of Abraham Gesner*. Hantsport NS: Lancelot, 1995.

Neale, Gladys E. *Made in Canada: We Got There First*. Saint John: Laubach Literacy of Canada, 1994.

Quinpool, John. *First Things in Acadia*. Halifax: First Things Publishers, 1936.

Ripper, William C.H. *Steam-Engine Theory and Practice*. London: Longmans, Green, 1927.

Rodger, Glenn B. and David G. Taylor. *Fredericton: A Postcard Trip to the Past*. Fredericton: Atlantex Publications, 1998.

Solway, Kenneth. *The Story of the Chestnut Canoe*. Halifax: Nimbus, 1997.

Soucoup, Dan. *Maritime Firsts: Historic Events, Inventions and Achievements*. East Lawrencetown NS: Pottersfield, 1996.

Walker, Albert W. *Text Book of the Patent Laws of the United States of America*. 4th ed. New York: Baker, Voorhis, 1904.

Web sites:
http://www.stanfields.com/history.html
http://www.hasbroscrabble.com/scrabble101.html
http://www.fullerbrush.com/
http://www.ganong.com/
http://www.italbusiness.it/snowbike/
http://www.uspto.gov/
http://strategis.ic.gc.ca/sc_mrksv/cipo/

INDEX

A

Airplane Position Indicator 67

Allin, William Bradford 49

Amherst NS 83

Antigonish NS 67

Austin, Henry T. 26, 84

Automobile Backup Light 70

Automobile parts 60, 68, 70, 83, 84

B

Baddeck NS 32

Baggage Check, Separable 63

Barker, William Henry 85

Barnes, Charles C. 53

Bateman, George Charles 85

Bathurst NB 45

Bear River NS 54

Beaverbrook, William Maxwell Aitken, Lord 68

Bell, Alexander Graham 32

Bertrand, Philias 78

Berwick NS 19

Black, James William 19

Blacks Harbour NB 26, 84

Boat Propeller 58

Boiestown NB 79

Boston MA 32, 58

Boutilier, Walter H. 27

Bryce, Frank W. 27

Bustin, Robert 85

C

Campbell, Thomas 33

Canadian Standards Association 74

Canoe, Wood and Canvas 65

Cans, Key-Opening 21, 26, 84

Cantelo, John Solomon 49

Cap with Ear and Head Band 36

Charlottetown PEI 49, 75

Chesley Corner NS 56

Chestnut Canoe Company 65

Chestnut, Harry 65

Chestnut, William T. 65

Chicken Bone 20

Chocolate Nut Bar 20

Chocolate Peppermint Rolls 20

Clark, Joseph Sutton 21

Clothes Washer with Wringer Rolls 28

Coleman, Stephen Leonard Chauncey 68, 83, 84

Confectionary Marker 20

Connecting Link 55

Connors Bros. Ltd. 26

Connors, Lewis 84

Coulson, James C. 36

Crossword Game 41

Crowell, W.A. 85

Cultivating and Hilling Machine 57

Cycle Runner 64

D

Daine, Richard 85

Dalhousie NB 61

Daman, L.W. 80

Dartmouth NS 29, 76, 85, 86

Davy, Arthur 66

Deer Island NB 23

Dominion Steel and Coal Corporation 74

Dump Box for Truck 71

E

Egerton NS 50

Elliott, James 46

F

Faucets, Combined Hot and Cold Water 33
Fenerty, Edward Lawson 85
Fish, Frozen Marinated 27
Fish Oil Products 23
Fishway 45
Forbes, John 29, 76, 85
Fredericton NB 47, 62, 65, 68, 83, 84
Fuller, Alfred C. 37, 87
Fuller Brush Company 37, 38, 87-88
Furnace, Pipeless 80-81

G

Gallagher, Bernard 85
Ganong Bros. Ltd. 20, 23
Ganong, Gilbert W. 20
Ganong, Whidden 23
Garments 35, 36, 39, 84
Gay, Robert 85
Gesner, Abraham 48
Gill, Abraham 51
Goodenough, Sidney Herbert 86
Goudge, Henry 85
Grand Pré NS 39
Grand River PEI 49
Grant, Charles L. 39
Great Bend KS 38

H

Halifax NS 27, 29, 48, 70, 85, 86
Hall, Arthur Wesley 73

Harris, Robert Carr 61
Harrison, Frank A. 86
Harrison, Thomas 85
Hartford CT 37
Hasbro Inc. 42
Hay Carrier 51-52
Holman, Robert C. 57
Horsford, Samuel 85
Household Cleaning Products 28, 30, 31, 87-88
Hudson's Bay Company 65
Hughes, Joseph Stuart 56

I

Ice Cream Soda 19
Ironing Board 31

J

Jenkins, George 49

K

Kentville NS 22
Kerosene 48

L

Larkin, John T. 85
Leadbetter, George W. 40
Lewis, Charles 84
Lipsett, James Thomas 77
Lollipops 20
Lyons, John Mitchell 63

M

Mabou NS 60
MacGillivray, John D. 67

Mackie, Irwin Cameron 74
Mawhinney, Robert T. 71
McAvity, Alexander 46
McCloskey, Lawrence St. Clair 79
McColl, Francis Partridge 84
McDonald, Edward R. 41
McFarlan, Richard 45
McKeen, Samuel 60
McNaughton, Isaac 50
McVay, William Alexander 73
Metal Protection 86
Miller, James H. 62
Miller's Flanger 62
Moncton NB 63
Montreal QC 27
Munro, Donald 55
Murray Harbour PEI 49

N

Negus, Frank 86
New Glasgow NS 66

O

Odometer 60
Oxford NS 85

P

Parson, Osborne H. 86
Patch, John 58
Pavement, Red 75
Phillips, Henry James 75
Pictou NS 55

Pool Cue Tip,
   Replaceable   40
Poole, John Henry   83
Porter, John B.   31
Potato Digger   49
Propeller, Variable Pitch
   72
Pulleys, Rubber-Lined
   50

Q

Quick-Release Buckle
   66

R

Rand, Rupert Leslie   83
Rein Connector   66
Reynolds, W.E.   49
Robertson, John R.   86
Roller Bearing   54
Rooney, James   22
Ross, James A.   70
Rotary Ventilator   77
Rothesay NB   72, 86
Round Hill NS   86

S

Sackville NB   53, 80, 86
Saint Hilaire NB   82
Saint John NB   28, 30,
   33, 34, 46, 71, 77, 78,
   83, 85
Sardine Cans   26, 84
Saw Tooth, Inserted   78
Scrabble   42
Scuba Tank   46
Sederquest, J. Roswell   64
Shediac NB   41

Shoes, Gum Rubber   39
Sidewalk, Partitioned
   Concrete   73
Skates, Clip-On   29
Skates, Steel   29, 85-86
Smith, Chesley Ernest   83
Snow Blower   61
Snowbike   64
Soap, Cold Water   30
Somerville MA   37
Spencer, John M.   83
Springhill NS   40
St. Andrews NB   84
St. George NB   21
St. Stephen NB   20, 64,
   73
Stabilizing Bar for
   Vehicle Suspension
   System   68-69
Stanfield, Frank   35
Stanfield, John   35
Stanfield's Limited   35
Starr Manufacturing
   Company   29
Steam Engine,
   Compound   47
Steel Rails, Tempering of
   74
Stevens, John   69
Stewart, Andrew James
   30
Stewart, Thomas McAvity
   34
Stiggins, William   49
Strain Gauges   76
Summerside PEI   57
Sydney NS   74

T

Taylor, Peter   86
Tea and Coffee Pot   22
Telephony, Electric   32
Thermal Windowpane
   79
Therriault, Joseph Leon
   82
Thomas, George Welton
   54
Thompson, Joseph   83
Thompson, Rachel A.   85
Tibbets, Benjamin F.   47
Tidal Motors   86
Toilet Bowl, Vortex-
   Flushing   34
Toothbrush with
   Replaceable Bristles
   37-38
Track Clearer   62
Truro NS   35, 36, 83, 84
Turnbull, John E.   28
Turnbull, W.R.   86
Turnbull, Wallace Rupert
   72, 86
Typewriter and
   Calculating
   Mechanisms   86

U

Underwear
   Combination,
   Adjustable   35
University of New
   Brunswick   69

V

Vane Pump   53

W

Washington DC 58, 76

Water Purification 86

Weatherstrip 82

Welsford NS 37

Wentworth, Hartley A.
   23

Westfield NB 83

Whelpley, Edwin H. 85

Whelpley, James Albert 85

Whitman, George 86

Windsor NS 85

Wood Pulp in Sheets 56

Wood Pulp Press 56

Y

Yarmouth NS 31, 58

Mario Theriault is a professional patent agent who lives in
Fredericton, New Brunswick. A graduate in industrial engineering
from Université de Moncton, he has more than twenty years of
experience in fields including electric power distribution, pulp
and paper, metal working, machine design and quality systems.
In 1995, he became a registered patent agent with his own firm,
Mario D. Theriault & Company. His passion for the patent
system and its value to humanity shows on every page
of *Great Maritime Inventions, 1833-1950*.
Visit his Website at http://www.patentway.com.